Madge made fudge for
the fair. Would the judge like
Madge's fudge? She wanted
to win a badge!

1

Madge made her fudge from scratch. She put a smidge of the batch in her mouth.

"Yuck! This batch is botched!" said Madge. "The fudge is too yucky."

So Madge made fudge once more. She made another batch from scratch.

"I won't botch this batch," she said. "I pledge to make better fudge!"

Madge tested the next batch. She nibbled a little smidge.

"Yum! Very fine!" said Madge. "No fudge can match this batch!"

Madge took her fudge to the fair. There were other fudge cooks there. Every cook looked at the judge.

"Fetch the fudge!" said the judge.

Every batch of fudge was
set on the ledge. The judge
would then sample the fudge.

He tested a smidge of each
batch. He was a great fudge
judge!

"Madge's fudge is tops," said the judge.

Madge got her badge!

"Every so often I botch a batch," said Madge. "But this batch of fudge is my best!"

"This is indeed very fine fudge!" said the judge. "Your fudge is top-notch!"

The End